SERVAL CATS

Tim Cooke

Grolier
an imprint of
SCHOLASTIC

www.scholastic.com/librarypublishing

Published 2009 by Grolier
An Imprint of Scholastic Library Publishing
Old Sherman Turnpike
Danbury, Connecticut 06816

For The Brown Reference Group
Project Editor: Jolyon Goddard
Picture Researchers: Clare Newman, Sophie Mortimer
Designer: Sarah Williams
Managing Editor: Tim Harris

Volume ISBN-13: 978-0-7172-8072-8
Volume ISBN-10: 0-7172-8072-1

Library of Congress Cataloging-in-Publication Data

Nature's children. Set 6.
p. cm.
Includes index.
ISBN-13: 978-0-7172-8085-8
ISBN-10: 0-7172-8085-3
1. Animals--Encyclopedias, Juvenile. I. Grolier (Firm)
QL49.N387 2009
590.3--dc22
2008014675

Printed and bound in China

PICTURE CREDITS

Front Cover: **Shutterstock**: Timothy Craig Lubcke.

Back Cover: **Shutterstock**: Kitch Bain, Eric Gevaert, David W. Hughes, Serg Zastavkin.

Corbis: Karl Ammann 21, Peter Johnson 34, Galen Rowell 26, Kennan Ward 46, Terry Whittaker 13, 42–43, 45; **NHPA**: Stephen Krasemann 18, Gerard Lacz 6, 38, Christophe Ratier 33, Kevin Schafer 9, 14, 17, 22, 30; **Shutterstock**: Kitch Bain 25, 37, Henk Bentlage 2–3, Patrick Hermans 5, Arnold John Labrentz 10, Kristian Seculic 29, Serg Zastavkin 4.

Contents

Fact File: Serval Cats 4
No Cheetah 7
Big and Small 8
Measuring Up 11
Jump to It! 12
Spots or Not 15
Serval Territory 16
Shrinking Wetlands 19
Night Creatures 20
A Good Day's Rest 23
Keep Out! 24
Water Babies 27
Noisy Animals 28
Striking a Pose 31
A Varied Diet 32
Listening for Food 35
A Skillful Hunter 36
We Have Lift Off 39

Living Alone 40
Finding a Mate 41
Feature Photo 42–43
The Kittens Arrive 44
Young Behavior 47
Under Threat 48

Words to Know 49
Find Out More 51
Index 52

FACT FILE: Serval Cats

Class	Mammals (Mammalia)
Order	Carnivores (Carnivora)
Family	Cat family (Felidae)
Genus	Serval cats, caracals, and African golden cats (*Caracal*)
Species	Serval cats (*Caracal serval*)
World distribution	Southern, western, and eastern Africa; a small population lives in northern Africa
Habitat	Live mainly in grassy plains and savanna, often near lakes or rivers
Distinctive physical characteristics	Extremely long legs in proportion to their body; large, disk-shaped ears; relatively short, banded tail; powerful jaws and sharp teeth; most servals have a tawny coat, covered in black spots
Habits	Usually solitary; hunt mainly at dusk
Diet	Small burrowing rodents, such as mole rats; birds, frogs, and fish; occasionally the young of larger animals, such as antelope

Introduction

Serval cats are small wild cats that are found in many places in Africa. Their long legs, small head, and large ears give them an unusual appearance. Like all types of cats, servals have an agile body. Their keen senses make them excellent hunters. But servals have to be careful because other hunters, such as leopards, may kill them. Servals' main enemy is humans, however. People turn the places where the cats live into farmland or kill them for their beautiful coat.

The large ears of a serval make it instantly recognizable.

A serval rests under a tree in the evening on the African savanna.

No Cheetah

As the sun goes down over the plains of Africa, something rustles in a clump of tall grass. A small, slender cat sticks up its head to look around as it prepares for an evening's hunt. Although this is a serval, you might think at first that you are looking at a baby cheetah. Servals and cheetahs have the same kind of coat—tawny colored with black spots.

So how can you tell that you are looking at an adult serval and not a baby cheetah? For one thing, compared to the rest of its body, a serval's head looks quite small—apart from its ears! Its ears are enormous and stick up from its head a little like radar dishes. Another difference is that, although the serval is smaller than the cheetah, it seems very tall. In fact, the serval has the longest legs of any of the cat family in proportion to its overall size. And while a cheetah has a long, sweeping tail, the serval's is short and sometimes sticks straight up in the air.

Big and Small

The cat family is divided into two groups. These are big cats and small cats. Big cats include lions, tigers, cheetahs, and leopards. The small cats include bobcats, ocelots, caracals, and servals. Despite their names, the differences between big and small cats are not only a matter of size. Larger small cats, such as cougars, can be as big as some of the smaller big cats, such as cheetahs.

One way to tell which group a cat belongs to is by the sounds it makes. Most of the big cats can roar, but they cannot purr continuously—both when breathing in and out. The small cats, on the other hand, can purr continuously, but they cannot roar.

All the small cats are related to one another and have similar habits, but they are **adapted** specially to suit their **habitats**. The serval lives in the African grasslands and is a master of hunting in the dense undergrowth.

The African black-footed cat is smaller than a house cat, but it is a fierce hunter.

A serval's coat helps it blend in with its surroundings, hiding it from prey and predators.

Measuring Up

The serval is one of the bigger small cats. The largest males weigh up to 40 pounds (18 kg) and up to 3 feet (1 m) long, with a tail of about 17 inches (35 cm). Female servals are usually slightly smaller than males. The serval's long legs make it appear very tall for its size. An adult is up to 2 feet (60 cm) tall at the shoulders. With such long legs and a long neck, it is easy for the serval to peer over the tall grass where it lives.

The serval has a beautiful yellow-brown coat, marked all over with black spots and stripes. Underneath, the serval's belly is nearly white, hiding it from **predators** and **prey**. Its tail has black rings and a black tip, as if it had been dipped in a paint can! Can you guess what all these blotches and marks are for? If you look at thickets and clumps of grass, you'll see many patterns of light and shadow. The serval's marks are **camouflage**—they help it blend in with the background.

Jump to It!

Servals have unusually long legs for cats. Compared with a pet cat, a serval looks as if it's wearing stilts! You can guess that those long legs make the serval a very fast hunter, though it can run only for short distances before it gets tired.

But can you guess what else those long, spindly legs might be good for? The serval is one of the cat family's champion high-jumpers. In tangled clumps of tall grass the best way for the serval to get around is in a series of high bounces, right over the top of anything that gets in the way. Long legs are also good for sprinting up tree trunks, either to chase prey or to escape from predators.

Like all cats, servals keep themselves very clean.

On dry savanna, servals have large, dark spots, like those of a cheetah, which help camouflage them.

Spots or Not

Depending on where they live, servals have a coat that suits them best. Where it's drier, in eastern and southern Africa, the serval's tawny coat has bold black spots and bars, a little like those of a cheetah or leopard. In other regions, however, servals don't have spots and bars. Instead, their coat is covered with brownish flecks—a little like freckles. The two forms look completely different from each other! Scientists used to think that the freckled servals were a completely different animal, which they called the servaline. Then they found out that servals and servalines could be brothers and sisters. Some servals, called **melanistic**, are black all over.

Serval Territory

Servals live in most parts of Africa, except for the deserts of the north and the thickest jungles. Most of all they live in great grassy plains called **savannas** or around the edge of forests, where it is easy for them to see and hear their prey. The tall grass hides them as they hunt. Usually, savannas are dotted with bushes or lightly wooded forests where the servals can take shelter from the hot sun during the day.

Servals particularly like living in damp places such as marshes and near rivers, where they hide in beds of reeds. Some servals also live high in the foothills of Africa's mountain ranges. In the mountains, there is plenty of open moorland to hunt in and forests or stands of bamboo to provide some shelter. A few servals can be found in the arid north of Africa, but they are in danger of dying out there.

A serval hunts in a marsh, poised to jump and pounce on its prey.

A thunderstorm over the African savanna signals the start of the rainy season.

Shrinking Wetlands

Like all animals on the savanna, the serval depends on the regular rainfall that sweeps the grasslands during the African winter.

At the height of the dry season, when the sun beats down on the plains, the grass dries out and turns golden. Water holes and marshes dry up. The areas of lush vegetation shrink. The servals find their watery habitat shrinking. As more grassland dries up, the wetlands that still exist are reduced to small patches, separated by dry areas. At this time, servals find it difficult to travel from one area to another.

That is the time of year when most servals **mate**. The **kittens** are born in the wet season, when the grasslands cover the largest area. There is plenty of food for the kittens' mother to hunt, enabling her to feed her new family.

Night Creatures

Most of the small cats are **nocturnal**. That means that they sleep most of the day and stay up all night to do their hunting. Why do you think they might live in such a way? Perhaps they hunt animals that are nocturnal, too, so they can find them only after dark.

It doesn't matter to the serval whether it hunts in the day or at night. Their sharp hearing and excellent eyesight mean that they can locate their prey even in the dead of night. The serval isn't truly nocturnal, though. Some of the animals it hunts are active during the day, so the serval starts hunting in the late afternoon, while it is still light. Near places where people live, however, the serval stays hidden all day and ventures out into the grassland only when night has fallen.

The serval's excellent eyesight allows it to locate its prey at night.

Servals like to rest in a thicket during the hottest part of the day.

A Good Day's Rest

The African savanna can be a tough place to take a nap. There are a lot of hunters around and not too many places to find good shelter. After a long night of finding food, servals like to spend the early morning stretched out on a rock or **termites' mound**. From this viewpoint they can keep a close watch on their **territory** while they catch some sun.

When they get drowsy, they find somewhere a little more private. Servals flatten down nests in clumps of tall grass or reeds where they curl up and sleep. It's very difficult to spot them once they're in their nest, because the spots on their coat help them blend into the grass. Even if a hunter does find a nest, it might not find a serval! The crafty cats move from place to place often and don't use the same home for very long.

Keep Out!

Like the other small cats, servals are **solitary** animals. They mark out their own hunting territory and get annoyed if another cat crosses the boundary. This is to make sure that there are never too many hunters chasing the same prey in a certain area. The serval does not get into fights because it marks its territory.

Servals patrol the boundaries of their territory, wearing well-trodden paths as they go. At key sites, they leave small deposits of dung where other servals will easily see or smell them. Servals also squirt their territory with urine, which is so smelly and greasy that it makes green plants wilt! In addition, servals use their **claws** to scratch marks on rocks, trees, and termites' mounds to tell other servals to keep out.

A serval patrols its hunting territory.

Servals love water and will often hunt for fish in rivers and water holes.

Water Babies

Everybody knows that cats don't like water—right? Wrong! The serval loves getting wet and is an expert swimmer. It splashes into rivers and water holes or into the muddy marshes, often close to where it lives. Why do you think servals like water so much? Perhaps it helps the cat cool down if the sun is too hot. Perhaps it helps get rid of pesky bugs and **ticks** from the serval's fur coat. Or perhaps the serval is looking for some of its favorite foods. The serval's long legs are really useful for reaching down into the water and scooping up a tasty-looking fish that gets too close. In a few very wet places, servals eat a lot of frogs.

Noisy Animals

The African savanna can be a noisy place. The wind rustles through the grass, animals gallop past noisily, birds call in the scattered trees. The servals have a whole range of voices to get themselves heard in all this noise. Like the other small cats, servals cannot roar. Instead, a serval often makes a high-pitched **chirp** that sounds more like a bird than a cat. It sounds a little like "how-how-how."

But that's not the only sound the serval makes. It meows like a house cat. It hisses when it gets mad, grunts when it gets excited, and growls and barks if it feels threatened. Sometimes, servals even pant like a dog.

Big cats, such as this lion, can make the savanna a noisy place with their roars.

A warning snarl from this serval tells intruders to stay back!

Striking a Pose

One of the ways servals communicate is with their body. They adopt exaggerated poses that carry all sorts of messages. If a serval meets an intruder in its territory, it arches its back and stretches its legs to make itself seem as tall as possible. The serval's fur **bristles**. All the while, the cat barks and growls and sometimes slashes at its rival with its front claws.

Another way servals give signals to one another is by raising and flattening their large ears. Their ears have a white patch on the back, ringed in black, making them easier for other servals to see. Tigers and leopards have similar spots on their ears.

The serval's tail gives the most distinctive signal, though. The serval raises it so that it points straight up in the air and then shivers it, as though it was vibrating. That is usually a friendly sign that a serval gives when it recognizes another serval.

A Varied Diet

Like all the cats, servals are **carnivores**, or meat eaters. They hunt mainly small **rodents**, hares, and birds, as well as fish and frogs. But a serval is not particularly fussy about what it eats. Snakes, lizards, termites, and grasshoppers are all on the menu at different times. Most cats sometimes eat grass, but the serval eats a lot of fruit, such as avocados and bananas. Servals are also agile climbers and can catch small tree-dwelling mammals, such as **hyraxes**. If a serval wants to catch a larger meal, such as a young antelope, it needs help and sometimes joins up with another serval to **ambush** its prey. Joint hunting is unusual, though, because the cats usually lead such a solitary life.

The African hare is a tasty meal for a serval, but it is quick and difficult to catch.

Mole rats are the best dish on the menu for servals, and servals are very good at finding them!

Listening for Food

Imagine if all your food was buried underground. How would you be able to find it? If you look in your backyard, you might see signs of creatures who live under the surface. These might include wormholes, molehills, or even fox **dens**. But how do you know where the worm, mole, or fox is? The mole rat is the serval's favorite food, and the serval has special weapons to find it—its ears. The serval's round, erect ears look and work almost like radar dishes. The hearing **organs** inside are specially sensitive and can even hear sounds too high-pitched for human ears. Amid the rustling grass and trees of the savanna, the serval picks out the sounds that mean food. If a mole rat or other rodent is **burrowing** underground or nibbling food somewhere, it had better be careful! The serval's fine ears can hear it, even from more than 20 feet (6 m) away!

A Skillful Hunter

The serval's long legs make it a good hunter. When it locates its prey, such as a mole rat coming out of its burrow, the serval crouches low on the ground and remains still until it makes sure its prey is as far from shelter as possible. Then the serval leaps high up in the air so that it lands on its unsuspecting victim with all four paws. At once, the cat follows up with a sharp bite. Its teeth have special **nerves** to feel for the soft gaps in its victim's backbone.

When a serval hears rustling in the long grass, it often bounces up and down while swiping down at the rustling place with its long front legs to hook its victim. Servals even kill snakes by slapping very hard at the snake's head with their outstretched paw. Not even a burrow can protect small animals from a hungry serval! The cat's long legs and claws can reach into holes and scratch away at the mounds in which rodents often hide.

This serval has caught sight of a tasty meal and slowly creeps forward.

A serval stands on its back legs, ready to leap into the air to catch an unsuspecting bird.

We Have Lift Off

One of the serval's favorite foods is birds such as quails, guinea fowl, bustards, and flamingos. If a bird is in a tree, the serval can leap into the air up to 10 feet (3 m) to knock it to the ground. Or if the bird has been hiding in the long grass and tries to fly away, the serval can leap to catch it in midair.

Like a pet cat, the serval likes to play with its dead or dying prey. It takes the dead bird in its mouth and throws it up in the air. It then stands up on its back legs to swipe at the bird as it falls. The cat also uses its mouth for plucking dead birds, carefully pulling out all the feathers before it settles down to its meal.

Living Alone

Like many cats, servals spend most of their life on their own. When the serval kittens are old enough to hunt for themselves, they leave their mother for good.

Females always find a territory of their own, but young males sometimes associate with their brothers. They hunt in pairs to capture large game such as antelope or impala, which are too big for females or young male servals to tackle alone. Sometimes even a serval hunting on its own might catch a large victim, such as a hare, which is too much for it to eat at once. When it is full, the serval carefully rakes grass and leaves over the remains of its meal to try to hide it from other predators.

Finding a Mate

During the driest times of the year, a female serval gets ready to find a partner. She has a special call to attract a male—a short, sharp meow that carries easily out over the grasslands. She calls in repeated, short bursts. If a male serval is nearby, he makes his way toward the sound. Instead of being as aggressive as she usually is to an intruder, the female meets the male with purring and rubs her cheeks against him, covering the male with saliva. Her tail sticks straight up in the air, vibrating with a shivering movement. Sometimes, she rolls on her back to show her bright white belly. That is a sign that she is ready to mate. When they have finished mating, the male serval does not stay with his new mate for long.

A pair of serval kittens waits for the return of their mother with food.

The Kittens Arrive

The female serval is pregnant for 9 to 11 weeks. She then searches out a sheltered place to give birth. This place is usually a spot in the middle of dense bushes, a hollow tree, a **crevice** in some rocks, or even the abandoned burrow of a porcupine or an aardvark.

The mother usually gives birth to two or three kittens at one time. Most servals are born either in March and April, in the spring, or in September and November, in fall.

The newborn kittens are blind and helpless and rely on their mother for milk during the first few weeks of their life. When they are born, their tail is very short, and their ears are small and folded over. The ears soon start to grow, however—quicker than any other part of the kitten's body!

A female serval with her kitten. Like most cats, a serval mother raises her family alone.

This serval kitten is quite young. Its eyes are still blue-gray.

Young Behavior

Serval kittens can be a handful for their mother. When they are only about three weeks old, they start trying to follow her every time she leaves the nest in search of food. The mother has to keep taking her kittens back to the nest. There, they lie down and make a chirping noise, like small birds.

When their mother leaves them alone, the kittens are smart enough to look after themselves. As soon as they see anything unusual, they freeze, keeping their body absolutely still to try to blend in with the scenery. At the same time, they follow every movement with their eyes.

Although all servals are born with blue-gray eyes, the eyes soon change into different colors. Females have yellow-brown eyes, and males have eyes that are more yellow-gray.

Under Threat

Servals have few natural predators. Leopards and dogs sometimes hunt the cats, but they are not the main danger servals face. One type, or subspecies, of serval, which lives in Morocco and Algeria in northern Africa, is endangered, meaning that it might die out.

The chief threats to servals are changes in their habitat and hunting. Humans drain the wet grasslands on which the cat depends to create land for agriculture. These changes reduce the amount of space for servals and their prey. And servals themselves are often hunted. They provide a good source of meat for local people, and their fur coat is very valuable.

The serval is also a popular pet. Humans keep them at home like house cats. However, many other people think that it is cruel to keep a wild animal in **captivity**.

Words to Know

Adapted When an animal is especially suited to its habitat, lifestyle, or diet.

Ambush To catch prey by hiding and waiting, then pouncing when they pass by.

Bristles Raises the hair so it stands on end.

Burrowing Digging through the soil.

Camouflage Colors and patterns that help an animal blend in with its surroundings.

Captivity Living in a human's home or a zoo and not free to roam in the wild.

Carnivores Animals that feed mainly on the flesh of other animals.

Chirp A short, sharp sound like that made by a bird, an insect, or a serval.

Claws The sharp, pointed nail-like growths at the ends of a cat's toes.

Crevice A crack in a large rock.

Dens The underground homes of animals such as foxes.

Habitats The types of places in which animals or plants naturally live.

Hyraxes	Small furry mammals that live in Africa and western Asia.
Kittens	Young cats.
Mate	To come together to produce young.
Melanistic	Having dark or black fur or skin.
Nerves	Cells in animals that carry messages about the surroundings to the brain.
Nocturnal	Active only at night.
Organs	Parts of the body such as the eyes, lungs, and liver.
Predators	Animals that hunt other animals.
Prey	Animals hunted by other animals.
Rodents	Small mammals with gnawing teeth. Mice, rats, and moles are rodents.
Savannas	Hot, grassy plains with little rainfall and few trees.
Solitary	Living alone.
Termites' mound	The small hill that is the home of antlike insects called termites.
Territory	An area that an animal defends from others of its own kind.
Ticks	Small insectlike animals that feed on the blood of larger animals.

Find Out More

Books

Bonar, S. *Small Wildcats*. New York: Franklin Watts, 2003.

Cat. DK Eyewitness Books. New York: DK Publishing Inc., 2004.

Web sites

Great Cats: Serval
nationalzoo.si.edu/Animals/GreatCats/servalfacts.cfm
A fact sheet on servals from the National Zoo.

Serval
www.enchantedlearning.com/subjects/mammals/cats/serval/printout.shtml
A printout and facts about servals.

Index

A, B, C
African black-footed cat 9
ambushing 32
big cats 8, 29
birth 44
biting 36
black servals 15
bobcat 8
calling 28
camouflage 11, 14, 23
caracal 8
cheetah 7, 8, 14, 15
claws 24, 31, 36
cleaning 13
climbing 32
coat 5, 7, 10, 11, 15, 23, 27, 31
communication 28, 31
cougar 8

D, E, F
dogs 48
ears 5, 7, 31, 35, 44
eyes 46, 47
eyesight 20, 21
fighting 24, 31
fur *see* coat

H
habitat 8, 16, 19
head 5, 7
hearing 20, 35
hunting 5, 7, 8, 10, 12, 16, 17, 19, 20, 24, 26, 27, 32, 36, 37, 40

J, K, L
jumping 12, 36, 38, 39
kittens 19, 40, 42, 44, 45, 46, 47
legs 5, 7, 11, 12, 27, 31, 36, 38, 39
leopard 5, 8, 15, 31, 48
lion 8, 29

M, N, O
mating 19, 41
milk 44
mole rat 34, 35, 36
mouth 39
neck 11
nest 23, 47
noises, making 28, 41, 47
ocelot 8

P, R
paws 36
play 39
predators 10, 11, 12, 40, 48
pregnancy 44
purring 8, 41
resting 6, 22
roaring 8, 28
running 12

S, T, W
scent-marking 24
sleep 20
small cats 5, 8, 11, 20, 24, 28
snarling 30
swimming 27
tail 7, 11, 31, 41, 44
teeth 36
territory 23, 24, 25, 31, 40
tiger 8, 31
weight 11